ANIMALS OF THE TUNDRA

SWIFT SNOWY OWLS

THERESA EMMINIZER

PowerKiDS press
New York

Published in 2022 by The Rosen Publishing Group, Inc.
29 East 21st Street, New York, NY 10010

First Edition

Portions of this work were originally authored by Roman Patrick and published as *Snowy Owls*. All new material in this edition authored by Theresa Emminizer.

Editor: Jill Keppeler
Book Design: Michael Flynn

Photo Credits: Cover, p. 1 Corey Hardcastle/500px Prime; (series background) Stefan Sorean/Shutterstock.com; p. 5 pchoui/iStock/Getty Images; p. 7 Javier Piva Flos/500px/Getty Images; p. 9 David G. Hemmings/Moment Open/Getty Images; p. 11 Bruce Lichtenberger/Photolibrary/Getty Images; p. 13 Bobo/500px/Getty Images; p. 15 Vicki Jauron, Babylon and Beyond Photography/Moment/Getty Images; p. 17 Riccardo Savi/The Image Bank/Getty Images; p. 19 Winfried Wisniewski/The Image Bank/Getty Images; p. 21 Jim H Walling/Shutterstock.com.

Library of Congress Cataloging-in-Publication Data

Names: Emminizer, Theresa, author.
Title: Swift snowy owls / Theresa Emminizer.
Description: New York : PowerKids Press, [2022] | Series: Animals of the tundra | Includes index.
Identifiers: LCCN 2020021704 | ISBN 9781725326514 (library binding) | ISBN 9781725326491 (paperback) | ISBN 9781725326507 (6 pack)
Subjects: LCSH: Snowy owl–Juvenile literature. | Tundra animals–Juvenile literature.
Classification: LCC QL696.S83 E46 2022 | DDC 598.9/7–dc23
LC record available at https://lccn.loc.gov/2020021704

Manufactured in the United States of America

CONTENTS

Snowy White

Snowy owls are as white as fresh snow! They live in the Arctic **tundra**. In the winter, sometimes they visit other places. You might see them in Europe, Asia, or North America.

Brown Beginning

Baby snowy owls have fluffy brown feathers! They'll become white as adults. **Male** owls get even whiter as they get older. **Females** keep darker stripes, or bars, on their feathers. The whitest owls are always males.

Warm in the Tundra

Snowy owls have many feathers. These feathers make them one of the heaviest owl species, or kinds! The feathers keep them warm in the cold tundra. Even their feet, beaks, and toes are covered with feathers.

What's for Dinner?

Snowy owls are good hunters. These owls eat small **mammals**, birds, and fish. Their favorite food is the lemming, a small, mouselike mammal. A snowy owl can eat more than 1,600 lemmings in one year!

Great Senses

Snowy owls have good eyesight, but they can't see under snow. They use their great hearing to listen for their dinner. They fly close to the ground and use their sharp claws to grab it!

Daytime Hunters

Snowy owls hunt during the day. The daytime can last for a very long time in the Arctic summer! These owls will hunt all day, but they move around the most at sunrise and sunset.

Teamwork

Mother owls build nests on the ground. They lay about three to eleven eggs. The father owl brings food to the mother as she keeps the eggs warm. Snowy owl **mates** often stay together for life.

Hatching!

When the eggs **hatch**, the father brings food to the babies. Mother owls may lay eggs at different times. So, they may still be warming some eggs after some have hatched.

Trying Their Wings

Baby owls leave the nest in less than a month. They can fly well by the time they are seven weeks old. Their parents keep feeding them for a few more weeks. In time, they'll go out on their own!

Fancy Features

GLOSSARY

female: A girl.

hatch: To break open or come out of.

male: A boy.

mammal: Any warm-blooded animal whose babies drink milk and whose body is covered with hair or fur.

mate: One of two animals that come together to produce babies.

tundra: Cold northern lands that lack forests and have permanently frozen soil below the surface.

FOR MORE INFORMATION

WEBSITES

Audubon: Snowy Owl
www.audubon.org/field-guide/bird/snowy-owl
Learn more about the snowy owl in the Audubon Society's guide to North American birds.

Snowy Owl
kids.nationalgeographic.com/animals/birds/snowy-owl/
National Geographic Kids presents this website full of facts about snowy owls.

BOOKS

Jaycox, Jaclyn. *Snowy Owls Are Awesome*. North Mankato, MN: Pebble Books, 2020.

Santos, Rita. *Life at the Poles: Snowy Owls*. New York, NY: Enslow Publishing, 2020.

INDEX